The Formulations In Space And Time

Ian Beardsley

2022

ENTITY BOOKS

Abstract: We show the protons are of a mathematical nature in space and time, as well as the elements in the periodic table of the elements, the hydrocarbons which are the skeletons of life chemistry, the radius of the solar system, and the abundance of the primordial elements hydrogen and helium, among many other structural elements connected to all of these.

Contents

1.0 Space And Time

We live in a 3 dimensional space. It requires 3 values to specify a point in space such as the point (1, 1, 1).

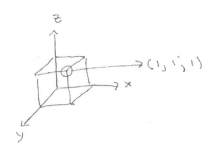

We can only draw the 3 dimensional shadow of a four dimensional hypercube because we cannot find a fourth right angle in our mind or space.

But we can draw a third right angle to a plane for which we see is the 2 dimensional shadow of a three dimensional cube.

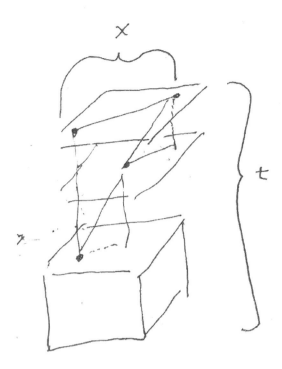

And as we move through a distance x we see we see we have travelled through a time t.

In order to specify a point P in space we need three values See illustration on page 7:

$$(x, y, z) = (a, b, c)$$

And, in order to do operations on how these three values may vary we need three unit vectors to specify their directions, each specified by three values two distinctly 0:

$$\vec{i} = (1,0,0)$$

$$\vec{j} = (0,1,0)$$

$$\vec{k} = (0,0,1)$$

And three distinct operators (such as $\dfrac{\partial}{\partial x}, \dfrac{\partial}{\partial y}, \dfrac{\partial}{\partial z}$. We define ∇ (nabla) and call it the del operator:

$$\nabla = \vec{i}\,\frac{\partial}{\partial x} + \vec{j}\,\frac{\partial}{\partial y} + \vec{k}\,\frac{\partial}{\partial z}$$

If we have a vector field $\vec{A}$, Then, where in a two dimensional plane we had one kind of multiplication, in three dimensions we we have two kinds of multiplication the dot product and the cross product:

$$\nabla \cdot \vec{A} = \frac{\partial \vec{A_x}}{\partial x} + \frac{\partial \vec{A_y}}{\partial y} + \frac{\partial \vec{A_z}}{\partial z}$$

Is a scalar. This, the dot product is also called the scalar product.

$$\nabla \times \vec{A} = \begin{pmatrix} \vec{i} & \vec{j} & \vec{k} \\ \dfrac{\partial}{\partial x} & \dfrac{\partial y}{\partial y} & \dfrac{\partial}{\partial z} \\ A_x & A_y & A_z \end{pmatrix}$$

Is a vector.

We don't have to cross nabla with $\vec{A}$, but we can cross $\vec{A}$ with another vector $\vec{B}$ as well dot them:

$$\vec{A} = A_1\vec{i} + A_2\vec{j} + A_3\vec{k}$$

$$\vec{B} = B_1\vec{i} + B_2\vec{j} + B_3\vec{k}$$

$$\vec{A} \cdot \vec{B} = A_1B_1 + A_2B_2 + A_3B_3$$

Or we can cross $\vec{A}$ with $\vec{B}$:

$$\vec{A} \times \vec{B} = \begin{pmatrix} \vec{i} & \vec{j} & \vec{k} \\ A_1 & A_2 & A_3 \\ B_1 & B_2 & B_3 \end{pmatrix} =$$

$$(A_2B_3 - A_3B_2)\vec{i} - (A_3B_1 - A_1B_3)\vec{j} + (A_1B_2 - A_2B_1)\vec{k}$$

In this instance we have three components each consisting of two terms for a total of six distinct terms, or products. They form a three by three matrix is $3^2 = 9$ elements. The key point I would like to make here is that there are six terms as there are six faces on a cube, and 2 times 9 is 18 is 6 times 3. The periodic table of the elements is periodic over 18 groups, each group of elements having related properties like number of valence electrons. For example carbon is in group 14 and has 18-14=4 valence electrons and silicon is in group 14 as well, below it in period 3 with 18-14=4 valence electrons as well. Regardless of whether we are working in rectangular coordinates as we have done so far, or some other coordinate system, like cylindrical or spherical, for which we need three values to specify a point as well. Thus we call it $\mathbb{R}^3$. Where $\mathbb{R}$ is the real numbers. However, we see cylindrical coordinates, which are employed for doing computations about straight lines like for the electric field that arises due to electrons moving in a wire, because as we progress along the straight line all distances from it are the same, the radius r of a cylinder, that it is contained int the rectangular coordinates because r is measured from the z-axis, θ is measured from the x-axis, and z is measured up from the x-y plane. Similarly, spherical coordinates are used for doing computations with spherical symmetry, like for the electric charge that arises from a charge because the electric field arises radially in all directions equally, so r is the same in all directions as well, and spherical coordinates are as well contained in the rectangular coordinate system because r is measured from the intersection of the of the x,y, and z axes. θ is measured from the x-axis and ϕ is measured from the z-axis. We see in fact that (See illustration next page, $\mathbb{R}^3$):

Cylindrical Coordinates

$$x = r\cos\theta, \ y = r\sin\theta, \ z=z, \ r^2 = x^2 + y^2, \tan\theta = y/x$$

Spherical coordinates

$$x = r\sin\phi\cos\theta, y = r\sin\phi\sin\theta, z = r\cos\phi, r^2 = x^2 + y^2 + z^2$$

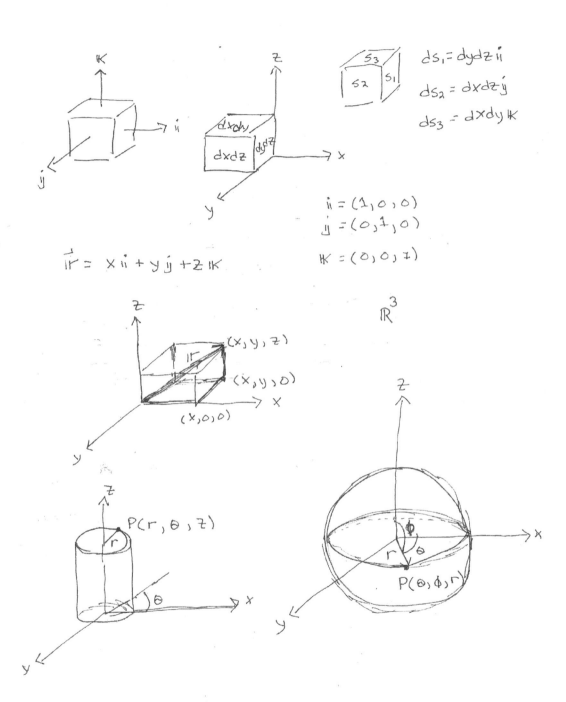

$$ds_1 = dydz\, \mathbb{i}$$
$$ds_2 = dxdz\, \mathbb{j}$$
$$ds_3 = dxdy\, \mathbb{k}$$

$$\mathbb{i} = (1,0,0)$$
$$\mathbb{j} = (0,1,0)$$
$$\mathbb{k} = (0,0,1)$$

$$\vec{r} = x\,\mathbb{i} + y\,\mathbb{j} + z\,\mathbb{k}$$

$$\mathbb{R}^3$$

(x,y,z)

$(x,y,0)$

$(x,0,0)$

$P(r,\theta,z)$

$P(\theta,\phi,r)$

2.0 The Periodic Table of The Elements

We wish to establish a theorem from all that has been said here.

Theorem 1.0 The Periodic Table of the Elements employs six fold symmetry. The periods of the periodic table are periodic over 18 groups, which is 3 times 6, the three being the components of $\mathbb{R}^3$, the number of values needed to specify a point in such a space.

	13	14	15
2	B		
3	Al	Si	P
4	Ga	Ge	As

Above we see the artificial intelligence (AI) elements pulled out of the periodic table of the elements. As you see we can make a 3 by 3 matrix of them and an AI periodic table. Silicon and germanium are in group 14 meaning they have 4 valence electrons and want 4 for more to attain noble gas electron configuration. If we dope Si with B from group 13 it gets three of the four electrons and thus has a deficiency becoming positive type silicon and thus conducts. If we dope the Si with P from group 15 it has an extra electron and thus conducts as well. If we join the two types of silicon we have a semiconductor for making diodes and transistors from which we can make logic circuits for AI.

As you can see doping agents As and Ga are on either side of Ge, and doping agent P is to the right of Si but doping agent B is not directly to the left, aluminum Al is. This becomes important. I call (As-Ga) the differential across Ge, and (P-Al) the differential across Si and call Al a dummy in the differential because boron B is actually used to make positive type silicon.

That the AI elements make a three by three matrix they can be organized with the letter E with subscripts that tell what element it is and it properties, I have done this:

$$\begin{pmatrix} E_{13} & E_{14} & E_{15} \\ E_{23} & E_{24} & E_{25} \\ E_{33} & E_{34} & E_{35} \end{pmatrix}$$

Thus E24 is in the second row and has 4 valence electrons making it silicon (Si), E14 is in the first row and has 4 valence electrons making it carbon (C). I believe that the AI elements can be organized in a 3 by 3 matrix makes them pivotal to structure in the Universe because we live in three dimensional space so the mechanics of the realm we experience are described by such a matrix, for example the cross product. Hence this paper where I show AI and biological life are mathematical constructs and described in terms of one another.

We see, if we include the two biological elements in the matrix (E14) and and (E15) which are carbon and nitrogen respectively, there is every reason to proceed with this reasoning if the idea is to show not only are the AI elements and biological elements mathematical constructs, they are described in terms of one another. We see this because the first row is (B, C, N) and these happen to be the only elements that are not core AI elements in the matrix, except boron (B) which is out of place, and aluminum (Al) as we will see if a dummy representative makes for a mathematical construct, the harmonic mean. Which means we have proved our case because the first row if we take the cross product between the second and third rows are, its respective unit vectors for the components meaning they describe them:

$$\vec{A} = (Al, Si, P)$$

$$\vec{B} = (Ga, Ge, As)$$

$$\vec{A} \times \vec{B} = \begin{pmatrix} \hat{B} & \hat{C} & \hat{N} \\ Al & Si & P \\ Ga & Ge & As \end{pmatrix} = (Si \cdot As - P \cdot Ge)\hat{B} + (P \cdot Ga - Al \cdot As)\hat{C} + (Al \cdot Ge - Si \cdot Ga)\hat{N}$$

$$\vec{A} \times \vec{B} = -145\hat{B} + 138\hat{C} + 1.3924\hat{N}$$

$$A = \sqrt{26.98^2 + 28.09^2 + 30.97^2} = 50 g/mol$$

$$B = \sqrt{69.72^2 + 72.64^2 + 74.92^2} = 126 g/mol$$

$$\vec{A} \cdot \vec{B} = AB\cos\theta$$

$$\cos\theta = \frac{6241}{6300} = 0.99$$

$$\theta = 8°$$

$$\vec{A} \times \vec{B} = AB\sin\theta = (50)(126)\sin 8° = 877.79$$

$$\sqrt{877.79} = 29.6 g/mol \approx Si = 28.09 g/mol$$

And silicon (Si) is at the center of our AI periodic table of the elements. We see the biological elements C and N being the unit vectors are multiplied by the AI elements, meaning they describe them! But we have to ask; Why does the first row have boron in it which is not a core biological element, but is a core AI element? The answer is that boron is the one AI element that is out of place, that is, aluminum is in its place. But we see this has a dynamic function.

The primary elements of artificial intelligence (AI) used to make diodes and transistors, silicon (Si) and germanium (Ge) doped with boron (B) and phosphorus (P) or gallium (Ga) and arsenic (As) have an asymmetry due to boron. Silicon and germanium are in group 14 like carbon (C) and as such have 4 valence electrons. Thus to have positive type silicon and germanium, they need doping agents from group 13 (three valence electrons) like boron and gallium, and to have negative type silicon and germanium they need doping agents from group 15 like phosphorus and arsenic. But where gallium and arsenic are in the same period as germanium, boron is in a different period than silicon (period 2) while phosphorus is not (period 3). Thus aluminum (Al) is in boron's place. This results in an interesting equation.

Equation 2.1 $$\frac{Si(As - Ga) + Ge(P - Al)}{SiGe} = \frac{2B}{Ge + Si}$$

The differential across germanium crossed with silicon plus the differential across silicon crossed with germanium normalized by the product between silicon and germanium is equal to the boron divided by the average between the germanium and the silicon. The equation has nearly 100% accuracy (note: using an older value for Ge here, it is now 72.64 but that makes the equation have a higher accuracy):

$$\frac{28.09(74.92 - 69.72) + 72.61(30.97 - 26.98)}{(28.09)(72.61)} = \frac{2(10.81)}{(72.61 + 28.09)}$$

$$0.213658912 = 0.21469712$$

$$\frac{0.213658912}{0.21469712} = 0.995$$

Due to an asymmetry in the periodic table of the elements due to boron we have the harmonic mean between the semiconductor elements (by molar mass):

Equation 2.2. $\dfrac{Si}{B}(As - Ga) + \dfrac{Ge}{B}(P - Al) = \dfrac{2SiGe}{Si + Ge}$

This is Stokes Theorem if we approximate the harmonic mean with the arithmetic mean:

$$\int_S (\nabla \times \vec{u}) \cdot d\vec{S} = \oint_C \vec{u} \cdot d\vec{r}$$

$$\int_0^1 \int_0^1 \left[\frac{Si}{B}(As - Ga) + \frac{Ge}{B}(P - Al) \right] dx\,dy \approx \frac{1}{Ge - Si} \int_{Si}^{Ge} x\,dx$$

We can make this into two integrals:

Equation 2.3

$$\int_0^1 \int_0^1 \frac{Si}{B}(As - Ga)dy\,dz \approx \frac{1}{3} \frac{1}{(Ge - Si)} \int_{Si}^{Ge} x\,dx$$

Equation 2.4

$$\int_0^1 \int_0^1 \frac{Ge}{B}(P - Al)dx\,dz \approx \frac{2}{3} \frac{1}{(Ge - Si)} \int_{Si}^{Ge} y\,dy$$

If in the equation (The accurate harmonic mean form):

Equation 2.5

$$\frac{Si}{B}(As - Ga) + \frac{Ge}{B}(P - Al) = \frac{Ge - Si}{\int_{Si}^{Ge} \frac{dx}{x}}$$

We make the approximation

Equation 2.6 $$\frac{2SiGe}{Si + Ge} \approx Ge - Si$$

Then the Stokes form of the equation becomes

Equation 2.7

$$\int_0^1 \int_0^1 \left[\frac{Si}{B}(As - Ga) + \frac{Ge}{B}(P - Al) \right] dydz = \int_{Si}^{Ge} dx$$

Thus we see for this approximation there are two integrals as well:

Equation 2.8 $$\int_0^1 \int_0^1 \frac{Si}{B}(As - Ga)dydz = \frac{1}{3}\int_{Si}^{Ge} dz$$

Equation 2.9 $$\int_0^1 \int_0^1 \frac{Ge}{B}(P - Al)dydz = \frac{2}{3}\int_{Si}^{Ge} dz$$

3.0 Proton-Seconds And The Vector Equilibrium

In order to present the elements as mathematical structures we need to explain the matter from which they are made as mathematical constructs. We need a theory for Inertia. I had found (Beardsley Essays In Cosmic Archaeology. 2021) where I suggested the idea of proton seconds, that is six proton-seconds, which is carbon the core element of biological life if we can figure out a reason to divide out the seconds. I found

Equation 3.1 $$\frac{1}{t_1\alpha^2 m_p}\sqrt{\frac{h4\pi r_p^2}{Gc}} = 6protons$$

Where h is Planck's constant 6.62607E−34 Js, r_p is the radius of a proton 0.833E-15m, G is the universal constant of gravitation 6.67408E-11 (Nm2)/(kg2), and c is the speed of light 299,792,459 m/s. And t_1 is t=1 second. α is the Sommerfeld constant (or fine structure constant) is 1/137. The mass of a proton is $m_p = 1.67262E - 27kg$.

Since plank's constant h is a measure of energy over time where space and time are concerned it must play a role. Of course the radius of a proton plays a role since squared and multiplied by 4π it is the surface area of our proton embedded in space. The gravitational constant is force produced per kilogram over a distance, thus it is a measure of how the surrounding space has an effect on the proton giving it inertia. The speed of light c has to play a role because it is the velocity at which event are separated through time. The mass of a proton has to play a role because it is a measurement of inertia itself. And alas the fine structure constant described the degree to which these factors have an effect. We see the inertia then in equation 3.1 is six protons over 1 second, by dimensional analysis.

The fine structure constant squared is the ratio of the potential energy of an electron in the first circular orbit to the energy given by the mass of an electron in the Bohr model times the speed of light squared:

$$\alpha^2 = \frac{U_e}{m_e c^2}$$

Matter is that which has inertia. This means it resists change in position with a force applied to it. The more of it, the more it resists a force. We understand this from experience, but what is matter that it has inertia?

I would like to answer this by considering matter in one of its simplest manifestations, the proton, a small sphere with a mass of 1.6726E-27 kg. This is a measure of its inertia.

I would like to suggest that matter, often a collection of these protons, is the three dimensional cross-section of a four dimensional hypersphere.

The way to visualize this is to take space as a two-dimensional plane and the proton as a two dimensional cross-section of a sphere, which would be a circle.

In this analogy we are suggesting a proton is a three dimensional bubble embedded in a two dimensional plane. As such there has to be a normal vector holding the higher dimensional sphere in a lower dimensional space. Thus if we apply a force to to the cross-section of the sphere in the plane there should be a force countering it proportional to the normal holding it in a lower dimensional universe. This counter force would be experienced as inertia. It may even induce in it an electric field, and we can see how it may do the same equal but opposite for the electron.

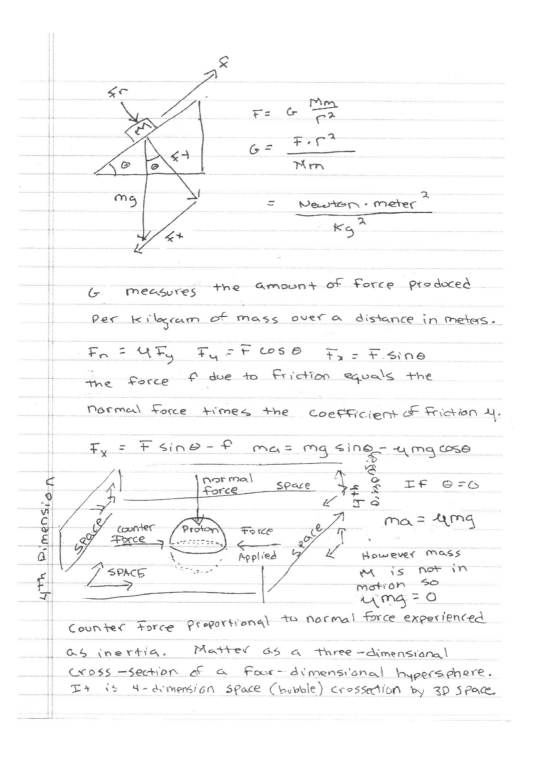

$$F = G \frac{Mm}{r^2}$$

$$G = \frac{F \cdot r^2}{Mm}$$

$$= \frac{Newton \cdot meter^2}{kg^2}$$

G measures the amount of force produced per kilogram of mass over a distance in meters.

$$F_n = \mu F_y \qquad F_y = F \cos\theta \qquad F_x = F \sin\theta$$

the force f due to friction equals the normal force times the coefficient of friction μ.

$$F_x = F \sin\theta - f \qquad ma = mg \sin\theta - \mu mg \cos\theta$$

If $\theta = 0$

$$ma = \mu mg$$

However mass M is not in motion so $\mu mg = 0$

Counter force proportional to normal force experienced as inertia. Matter as a three-dimensional cross-section of a four-dimensional hypersphere. It is 4-dimension space (bubble) crossection by 3D space.

Since we have

$$\frac{1}{\alpha^2 m_p} \sqrt{\frac{h 4\pi r_p^2}{Gc}} \int_{t_1}^{t_2} \frac{1}{t^2} dt = \mathbb{N}$$

Is a number of protons

$$\frac{1}{\alpha^2 m_p} \sqrt{\frac{h 4\pi r_p^2}{Gc}}$$

Is proton-seconds. Divide by time we have a number of protons because it is a mass divided by the mass of a proton. But these masses can be considered to cancel and leave pure number.

Generating A Table

We make a program that looks for close to whole number solutions so we can create a table of values for problem solving. I set it at decimal part equal to 0.25. You can choose how may values for t you want to try, and by what to increment them. Here are the results for incrementing by 0.25 seconds then 0.05 seconds. Constant to all of this is hydrogen and carbon. The smaller integer value of seconds gives carbon (6 protons at 1 second) and the largest integer value of seconds gives hydrogen (1 proton at six seconds) and outside of that for the other integer values of protons you get are at t>0 and t<1. Equation 1 really has some interesting properties. Here are two runs of the program(decpart is just me verifying that my boolean test was working right to sort out whole number solutions):

By what value would you like to increment?: 0.25
How many values would you like to calculate for t in equation 1 (no more than 100?): 100
24.1199 protons 0.250000 seconds 0.119904 decpart
12.0600 protons 0.500000 seconds 0.059952 decpart
8.0400 protons 0.750000 seconds 0.039968 decpart
6.0300 protons 1.000000 seconds 0.029976 decpart
4.0200 protons 1.500000 seconds 0.019984 decpart
3.0150 protons 2.000000 seconds 0.014988 decpart
2.1927 protons 2.750000 seconds 0.192718 decpart
2.0100 protons 3.000000 seconds 0.009992 decpart
1.2060 protons 5.000000 seconds 0.205995 decpart
1.1486 protons 5.250000 seconds 0.148567 decpart
1.0964 protons 5.500000 seconds 0.096359 decpart
1.0487 protons 5.750000 seconds 0.048691 decpart
1.0050 protons 6.000000 seconds 0.004996 decpart
0.2487 protons 24.250000 seconds 0.248659 decpart
0.2461 protons 24.500000 seconds 0.246121 decpart
0.2436 protons 24.750000 seconds 0.243635 decpart

By what value would you like to increment?: 0.05
How many values would you like to calculate for t in equation 1 (no more than 100?): 100
40.1998 protons 0.150000 seconds 0.199837 decpart
30.1499 protons 0.200000 seconds 0.149879 decpart
24.1199 protons 0.250000 seconds 0.119904 decpart

20.0999 protons 0.300000 seconds 0.099918 decpart
17.2285 protons 0.350000 seconds 0.228500 decpart
15.0749 protons 0.400000 seconds 0.074938 decpart
12.0599 protons 0.500000 seconds 0.059950 decpart
10.0500 protons 0.600000 seconds 0.049958 decpart
8.0400 protons 0.750000 seconds 0.039966 decpart
7.0941 protons 0.850000 seconds 0.094088 decpart
6.0300 protons 1.000000 seconds 0.029975 decpart
5.2435 protons 1.150000 seconds 0.243457 decpart
5.0250 protons 1.200000 seconds 0.024980 decpart
4.1586 protons 1.450000 seconds 0.158605 decpart
4.0200 protons 1.500000 seconds 0.019985 decpart
3.1737 protons 1.899999 seconds 0.173673 decpart
3.0923 protons 1.949999 seconds 0.092296 decpart
3.0150 protons 1.999999 seconds 0.014989 decpart
2.2333 protons 2.699999 seconds 0.233325 decpart
2.1927 protons 2.749999 seconds 0.192719 decpart
2.1536 protons 2.799999 seconds 0.153564 decpart
2.1158 protons 2.849998 seconds 0.115782 decpart
2.0793 protons 2.899998 seconds 0.079303 decpart
2.0441 protons 2.949998 seconds 0.044061 decpart
2.0100 protons 2.999998 seconds 0.009993 decpart
1.2433 protons 4.850000 seconds 0.243294 decpart
1.2306 protons 4.900001 seconds 0.230607 decpart
1.2182 protons 4.950001 seconds 0.218177 decpart

Here is the code for the program:

```c
#include <stdio.h>
#include <math.h>
int main(int argc, const char * argv[]) {

    int n;
    float value=0, increment,t=0, p=1.67262E-27, h=6.62607E-34,G=6.67408E-11,
c=299792459,protons[100],r=0.833E-15;

    do
    {
        printf("By what value would you like to increment?: ");
        scanf("%f", &increment);
    printf("How many values would you like to calculate for t in equation 1 (no more than 100?):
");
        scanf("%i", &n);
    }
        while (n>=101);
    {

        for (int i=0; i<n;i++)
          {
              protons[i]=((137*137)/(t*p))*sqrt(h*4*(3.14159)*(r*r)/(G*c));

              int intpart=(int)protons[i];
              float decpart=protons[i]-intpart;
```

```
t=t+increment;
if (decpart<0.25)
{ printf("%.4f protons %f seconds %f decpart \n", protons[i], t-increment, decpart);
}}}}
```

A very interesting thing here is looking at the values generated by the program, the smallest integer value, 1 second produces 6 protons (carbon) and the largest integer value 6 seconds produces one proton (hydrogen). Beyond six seconds you have fractional protons, and the rest of the elements heavier than carbon are formed by fractional seconds. These are the hydrocarbons the backbones biological chemistry.

Equation 3.2 $$\frac{1}{6\alpha^2 m_p}\sqrt{\frac{h4\pi r_p^2}{Gc}} = 1.004996352 seconds$$

$$\frac{1}{\alpha^2 m_p}\sqrt{\frac{h4\pi r_p^2}{Gc}} = (6protons)(1second)$$

$$\frac{1}{\alpha^2 m_p}\sqrt{\frac{h4\pi r_p^2}{Gc}} = (1proton)(6seconds)$$

$$\alpha^2 = \frac{U_e}{m_e c^2}$$

Thus we have six protons (carbon) is one second. We see 1 second predicts the radius of a proton,...

In that we have

$$\frac{1}{\alpha^2 m_p}\sqrt{\frac{h4\pi r_p^2}{Gc}}\int_{t_1}^{t_2}\frac{1}{t^2}dt = \mathbb{N}$$

And the periodic table of the elements is cyclical with 18 groups and

$$6 = \frac{1}{\alpha^2 m_p}\sqrt{\frac{h4\pi r_p^2}{Gc}}$$

Then perhaps we are supposed to write

$$\frac{3}{\alpha^2 m_p}\sqrt{\frac{h4\pi r_p^2}{Gc}}\int_{t_1}^{t_2}\frac{1}{t^2}dt = 18$$

In fact, what if the 3 is supposed to be pi, then

$$\frac{\pi}{\alpha^2 m_p}\sqrt{\frac{h4\pi r_p^2}{Gc}}\int_{t_1}^{t_2}\frac{1}{t^2}dt = 18$$

Then we would say that

k=18/pi=5.7229577951

The parameter in our constant with the most uncertainty is the radius of a proton r_p. If the 3 is supposed to be pi, then the radius of a proton becomes:

Equation 3.3
$$r_p = k\alpha^2 m_p\sqrt{\frac{Gc}{4\pi h}}$$

Which gives

$$r_p = 8.790587E - 16m$$

About 95% raw most recent value measured. But, if

$$\frac{1}{\alpha^2 m_p}\sqrt{\frac{h4\pi r_p^2}{Gc}}$$

Is supposed to be 6 and it is supposed to be multiplied by three to give 18 even which we need for chemistry so we have 18 protons in the last group of the periodic table which is important because we need argon with 18 protons for predicting valence numbers of elements in terms of their need to attain noble gas electron configuration. Then we get

$$r_p = 8.288587E - 16m = 0.829fm$$

This is in very close agreement with the most recent value measured which is

$$r_p = 0.833 + / - 0.014$$

We now formulate what I call Giordano's Relationship: Warren Giordano writes in his paper *The Fine Structure Constant And The Gravitational Constant: Keys To The Substance Of The Fabric Of Space*, March 21, 2019:

In 1980, the author had compiled a series of notes analyzing Einstein's geometric to kinematic equations, along with an observation that multiplying Planck's constant 'h' by '$1 + \alpha$', where 'α' is the Fine Structure Constant, and multiplying by 10^{23} yielded Newton's gravitational constant numerically, but neglecting any units.

Let's do that

(6.62607E-34Js)(1+1/137)(1E23)=6.6744E-11 Js

And it works, G is:

G=6.67408E-11 N(m2/kg2)

This is like saying gravitational constant masses equal ten to the 23 Planck velocities:

$$Gm_1 = 10^{23}h(1 + \alpha)(v_1)$$

$$Gm_1 = 10^{23}h(1 + \alpha)(v_1)$$

Where m1 is a kilogram and v1 is 1 meter per second. Avogadro's number is

$$6E23 atoms/mol = N_A$$

But, if we are working with kilograms we say

$$N_A = 6E26 atoms/mole$$

But H=0.001 kg/mol not 1.00 g/mol and carbon is C=0.012 kg/mol not 12.01 g/mol

Thus if H is the molar mass of hydrogen (1 g/mol=0.001kg/mol) then there are 6E26 atoms in a kilogram of it. We have

$$6 \cdot seconds \frac{kilograms}{meter} = 1000h \frac{N_A}{G}(1 + \alpha)$$

Remembering we now say in this system of units $N_A = 6E26 atoms/mol$ if we are working in kg.

But can we say

$$\frac{N_A}{H} = 6E23$$

And our equations become

$$6 \cdot seconds \frac{kilograms^2}{meter} = h \frac{N_A}{H} \frac{(1 + \alpha)}{G}$$

$$N \cdot kilograms^2 \frac{seconds}{meter} = h \frac{N_A}{E} \frac{(1 + \alpha)}{G}$$

Where E is any element and N is some number. Let us reformulate this as:

Equation 3.4 $$h \frac{(1 + \alpha)}{G} N_A H = 6.0003 \frac{kg^2 \cdot s}{m}$$

Where $N_A = 6.02E23$ and H=1 gram/atom

Because for hydrogen 1 proton is molar mass 1 gram, for carbon 6 protons is 6 grams and so on for 6E23 atoms per gram. Thus,...

$$N_A H = 6.02E23 \frac{atoms}{gram} \cdot \frac{1 gram}{atom} = 6.02E23$$

Since grams and atom cancel we can work in grams even though our equations are in kilograms. Let us not write H, since formally it is grams per mole of hydrogen but write

$$\mathbb{H} = 1 \frac{gram}{atom}$$

We have:

$$h \frac{(1+\alpha)}{G} \cdot N_A \mathbb{H} = 6.0003 kg^2 \cdot \frac{s}{m}$$

Or,...

Equation 3.5 $\qquad h(1+\alpha)N_A \mathbb{H} = 6Gx$

Where

$$x = 1.00 kg^2 \frac{s}{m}$$

Let us say we were to consider Any Element $\mathbb{E}$ say carbon $\mathbb{C}$. Then in general

Equation 3.6 $\qquad h \frac{(1+\alpha)}{G} \cdot N_A \mathbb{E} = 6.0003 kg^2 \cdot \frac{s}{m}$

We have

$$\mathbb{C} = \frac{6 grams}{6 protons} \quad \text{and } N_A = \frac{6(6E23 protons)}{6 grams}$$

Because there are six grams of protons in carbon which has 6 protons and 6 neutrons and a molar mass of 12. We have 12-6=6 grams of protons in the 12 grams of protons and neutrons. Thus

$$N_A \mathbb{C} = 6E23$$

And it follows that

$$h\frac{(1+\alpha)}{G} \cdot N_A \mathbb{C} = 6.0003 kg^2 \cdot \frac{s}{m}$$

We see in general since the atomic number Z is the number of protons in an atom that in general this holds for all elements $\mathbb{E}$ because

$$N_A = \frac{Z \cdot 6E23 protons}{Z \cdot grams}$$

And,

$$\mathbb{E} = \frac{Z \cdot grams}{Z \cdot protons}$$

Therefore we always have:

Equation 3.7 $\qquad N_A \cdot \mathbb{E} = 6E23$

Let us consider then equation 5. We multiply both sides by $\frac{r_p}{m_p^2}$.

$$\frac{r_p}{m_p^2} = 6.000 kg^2 \frac{s}{m} \cdot \frac{0.833E - 15m}{(1.67262E - 27kg)^2} = 1.7865E39s$$

That is we have

Equation 3.8 $\qquad \frac{r_p}{m_p^3} \cdot h\frac{(1+\alpha)}{Gx} \cdot N_A \mathbb{E} = 1.7865E39s$

We will return to this answer, but me we look at the significance of time in Quantum Mechanics first.

Thus returning to theorem 1.0 and considering equation 3.1 And 3.4

Theorem 1.0 The Periodic Table of the Elements employs six fold symmetry. The periods of the periodic table are periodic over 18 groups, which is 3 times 6, the three being the components of $\mathbb{R}^3$, the number of values needed to specific a point in such a space.

Equation 3.1 $\qquad \frac{1}{t_1\alpha^2 m_p}\sqrt{\frac{h4\pi r_p^2}{Gc}} = 6 protons$

Equation 3.4. $\qquad h\frac{(1+\alpha)}{G}N_A H = 6.000\frac{kg^2 \cdot s}{m}$

Thus in 3.1 we have 6 protons and in 3.4 we have $6\dfrac{kg^2 \cdot s}{m}$. Equation 1 is determined by $t_1 = 1s.$

Buckminster Fuller writes in Synergetics:

205.01 The geometrical model of energy configurations in synergetics is developed from a symmetrical cluster of spheres, in which each sphere is a model of a field of energy all of whose forces tend to coordinate themselves shuntingly or pulsatively, and only momentarily in positive or negative asymmetrical patterns relative to, but never congruent with, the eternality of the vector equilibrium…This closest packing of spheres in 60-degree angular relationships demonstrates a finite system in Universal geometry.

The NUMBER OF SPHERES ON ANY LAYER IS 10F+2
F = frequency, here F=1

Closest packing of equal radius sphers

VECTOR EQUILIBRIUM

Let us consider the twelve spheres packed around the central sphere to be the six protons and six neutrons of the carbon atom.

tetrahedron = 1

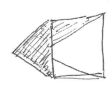 half
octahedron = 2

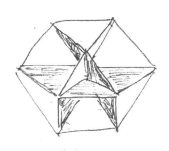

The volume of the
VECTOR EQUILIBRIUM
Is 8 tetrahedra
Plus six half octahedra
Equals tewenty tetrahedra

2 (6 Pyramids)(8 tetrahedra) = 1 icosahedron
= 20 tetrahedra

$$\frac{12}{8} = \frac{3 \cdot 4}{8} = \frac{3}{2} = 1\frac{1}{2} = 1.5$$

Volume of tetrahedron: $v = \frac{a^3}{6\sqrt{2}}$

Volume of octahedron: $v = \frac{\sqrt{2}}{3} a^3$

Volume of $\frac{1}{2}$ octahedron: $\frac{\sqrt{2}}{6} a^3$ (pyramid)

Volume of icosahedron: $v = \frac{5(3+\sqrt{5})}{12} a^3$

Thus we see carbon is 8 tetrahedra and 6 octahedra, each of which is 2 tetrahedra I found them precisely this in their two dimensional analogs the regular dodecahedron and regular octagon:

We see that the atomic radius of silicon the core element of artificial intelligence (transistor technology) fits together with the core element of biological life carbon if the silicon is taken as inscribed in a regular dodecagon and the carbon is taken as inscribed in a regular octagon. We have:

$$D = 1 + 2x, \quad x^2 + x^2 = 1^2, \quad 2x^2 = 1$$
$$2x^2 = 1, \quad x = \sqrt{2}/2, \quad D = 1 + \sqrt{2}$$
$$\text{Apothem: } a = (1 + \sqrt{2})/2 = 1.2071$$

For a regular dodecagon:

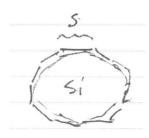

$$a = \frac{s/2}{tan(\theta/2)} = \frac{0.5}{tan(15°)} = 1.866$$

The radius of a silicon atom is Si=0.118nm and that of carbon is C=0.077nm:

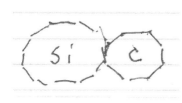

$$\frac{Si}{C} = \frac{0.118}{0.077} = 1.532$$

$$\frac{a_{Si}}{a_C} = \frac{1.866}{1.2071} = 1.54585$$

This has an accuracy $\frac{1.532}{1.54585}100 = 99\%$

But, we still have the question of 1 second in equation 3.1:

$$\frac{1}{t_1 \alpha^2 m_p} \sqrt{\frac{h 4\pi r_p^2}{Gc}} = 6protons$$

We see that the vector equilibrium employs 60-degree coordination. And that it involves a layer of 12 spheres. However we have a 12 hour daytime and and 12 hour night time, which is a 24 hour day (Earth rotates through 15 deg an hour, dodecagon). But there are 60 minutes in an hour and 60-degree coordination is connected to that in that the earth rotates through 360 degrees in 24 hours and 360/60 degrees is our six-fold symmetry of the periodic table. Each hour then of the 24 hours is divided by 60 minutes and each minute divided by 60 is 1 second.

We have our one second. Indeed I find the size of the solar system is connected to Silicon through its counterpart as a semiconductor material germanium. Let's look at that:

While we have

$$\frac{1}{\alpha^2 m_p} \sqrt{\frac{h 4\pi r_p^2}{Gc}} \int_{t_1}^{t_2} \frac{1}{t^2} dt = \mathbb{N}$$

Is a number of protons

$$\frac{1}{\alpha^2 m_p} \sqrt{\frac{h 4\pi r_p^2}{Gc}}$$

Is proton-seconds. Divide by time we have a number of protons because it is a mass divided by the mass of a proton. But these masses can be considered to cancel and leave pure number. We have that

$$6 \int_{\sqrt{2}}^{\sqrt{3}} \frac{1}{t^2} dt = 6 \left(\frac{1}{\sqrt{2}} - \frac{1}{\sqrt{3}} \right) = 0.78$$

$$\int_{\sqrt{2}}^{\sqrt{3}} cos^{-1}(x/2) dx = \frac{1}{6} \left(\sqrt{3}\pi - 6 \right) - \frac{\pi - 4}{2\sqrt{2}} = 0.21$$

Interestingly 78% is the percent of N2 in the atmosphere and 21% is the percent of O2 in the atmosphere (by volume). These are the primary constituents that make it up. The rest is primarily argon and CO2. This gives the molar mass of air as a mixture is:

$$0.78 N2 + 0.21 O2 = 29.0 g/mol$$

Now interestingly, I have found this connected to the solar system by considering a mixture of silicon, the primary constituent of the Earth crust, and germanium just below it in the periodic table, in the same proportions of 78% and 21%. Silicon is also the primary second generation semiconductor material (what we use today) and germanium is the primary first generation semiconductor material (what we used first). The silicon is directly below our carbon of one proton-second, silicon directly below that, and germanium directly below that, in the periodic table. So we are moving directly down the periodic table in group 14. The density of silicon is 2.33 g/cm3 and that of germanium is 5.323 g/cm3. Let us weight these densities with our 0.21 and 0.78:

$$0.21 Si + 0.78 Ge = 4.64124 g/cm^3$$

Now consider this the starting point for density of a thin disc decreasing linearly from the Sun to Pluto (49.5AU=7.4E14cm). Thus,…

$$\rho(r) = \rho_0 \left(1 - \frac{r}{R} \right)$$

Thus,…

$$M = \int_0^{2\pi} \int_0^R \rho_0 \left(1 - \frac{r}{R}\right) r\, dr\, d\theta$$

Or,…

$$M = \frac{\pi \rho_0 R^2}{3}$$

Thus,…

$$M = \frac{\pi (4.64124)(7.4E14)^2}{3} = 2.661E30$$

If we add up the masses of the planets in our solar system they are 2.668E30 grams.

Since

$$\frac{2.661}{2.668}(100) = 99.736$$

Meaning mixing germanium and silicon in the same proportion that occurs with N2 and O2 in the atmosphere and with

$$6\int_{\sqrt{2}}^{\sqrt{3}} \frac{1}{t^2} dt = 6\left(\frac{1}{\sqrt{2}} - \frac{1}{\sqrt{3}}\right) = 0.78$$

$$\int_{\sqrt{2}}^{\sqrt{3}} \cos^{-1}(x/2)\, dx = \frac{1}{6}\left(\sqrt{3}\pi - 6\right) - \frac{\pi - 4}{2\sqrt{2}} = 0.21$$

Where

$$6 = \frac{1}{\alpha^2 m_p} \sqrt{\frac{h 4\pi r_p^2}{Gc}}$$

In the first integral. See the following pages to see the computation of the mass of the planets in the solar system…

As we can see Jupiter carries the majority of the mass, Saturn a pretty large piece, and somewhat large Uranus and Neptune. We don't even list Pluto's mass. When we consider

Planet.	In Earth Mass
Mercury.	0.0553
Venus.	0.815
Earth.	1
Mars.	0.11
Jupiter.	317.8
Saturn.	95.2
Uranus.	14.6
Neptune.	17.2
Moon.	0.0123

Earth Mass In Grams: 5.972E27

Asteroid Belt: 4% of Moon (0.00492)

AU: 1.496E13 cm

Pluto (Edge of Solar System): 49.5 AU (7.4E14 cm)

Solar System (Mercury to Neptune): 2.6682E30 grams

$$6 \int_{\sqrt{2}}^{\sqrt{3}} \frac{1}{t^2} dt = 6 \left(\frac{1}{\sqrt{2}} - \frac{1}{\sqrt{3}} \right) = 0.78$$

$$\int_{\sqrt{2}}^{\sqrt{3}} cos^{-1}(x/2) dx = \frac{1}{6} \left(\sqrt{3}\pi - 6 \right) - \frac{\pi - 4}{2\sqrt{2}} = 0.21$$

We are considering $\sqrt{2}$ and $\sqrt{3}$. These come from

$$2cos(45°) = \sqrt{2}$$

$$2cos(30°) = \sqrt{3}$$

From 30 degrees to 45 degrees is 15 degrees. The Earth rotates through 360/24 is 15 degrees per hour. The hour is divided into 60 minutes and minutes into 60 seconds…

We have said

$$M = \frac{\pi \rho_0 R^2}{3}$$

For a thin disc. Thus we have a definition for the radius of the solar system, R_s:

Equation 3.9 $R_s = \sqrt{\dfrac{3M_p}{\pi(0.78Ge + 0.21Si)}}$

Where

Equation 3.10 $\dfrac{1}{\alpha^2 m_p}\sqrt{\dfrac{h4\pi r_p^2}{Gc}}\displaystyle\int_{\sqrt{2}}^{\sqrt{3}}\dfrac{1}{t^2}dt = 0.78$

Equation 3.111 $\displaystyle\int_{\sqrt{2}}^{\sqrt{3}}\cos^{-1}(x/2)dx = 0.21$

Equation 3.12 $air = 0.78N_2 + 0.21O_2$

Equation 3.13 $\dfrac{air}{H_2O} \approx \Phi$

M_p is the mass of all the planets. We also have that the molar mass of air to the molar mass of water is approximately the golden ratio. The interesting thing is we determine a definition for the radius of the solar system and predict the hydrocarbons (backbones of life) all in one fell swoop. Thus since if the universe is mostly hydrogen and helium if we can predict their relative abundances with proton-seconds then we can feel quite certain we are formulating things correctly, consider the following gaussian distribution illustrated...

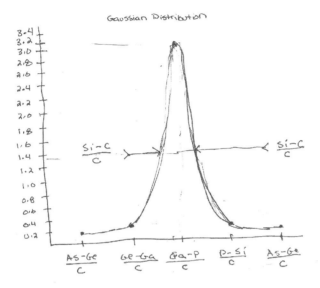

Gaussian Distribution

$$\frac{As-Ge}{C} = 0.18996$$

$$\frac{Ge-Ga}{C} = 0.2428612$$

$$\frac{Si-C}{C} = 1.33827$$

$$\frac{Ga-P}{C} = 3.22615$$

$$\frac{P-Si}{C} = 0.24051$$

$$\frac{As-Ge}{C} = 0.18996$$

$$\frac{As}{C} = \frac{74.9216}{12.011} = 6.23775 \qquad \frac{Ga}{C} = \frac{69.723}{12.011} = 6.04779$$

$$\frac{Ge}{C} = \frac{72.64}{12.011} = 5.8049288 \qquad \frac{P}{C} = \frac{30.9737620}{12.011} = 2.57878$$

$$\frac{Si}{C} = \frac{28.085}{12.011} = 2.33827$$

We consider a Gaussian wave-packet at t=0:

$$\psi(x,0) = Ae^{-\frac{x^2}{2d^2}}$$

We say that d which in quantum mechanics would be the delocalization length when squared is $\frac{Si-C}{C}$. A is the amplitude and we might say it is $\frac{Ga-P}{C}$. We write the wave packet as a Fourier transform which is:

$$\psi(x,0) = Ae^{-\frac{x^2}{2d^2}} = \int \frac{dp}{2\pi\hbar}\phi_p e^{\frac{i}{\hbar}px}$$

We use the identity that gives the integral of a quadratic:

$$\int_{-\infty}^{\infty} e^{-\alpha^2 x + \beta x}dx = \sqrt{\frac{\pi}{\alpha}}e^{\frac{\beta^2}{4\alpha}}$$

Solve the equation

$$i\hbar\partial_t\psi(x,t) = \frac{\hat{p}}{2m}\psi(x,t)$$

With the initial condition

$$\psi(x,0) = \int dp \cdot e^{\frac{p^2 d^2}{2\hbar^2}} \cdot e^{-\frac{i}{\hbar}px}$$

A plane wave is the solution:

$$e^{\frac{i}{\hbar}(px-\epsilon(p)t)}$$

Where, $\epsilon(p) = \frac{p^2}{2m}$

The wave-packet evolves with time as

$$\psi(x,t) = \int dp \cdot e^{\frac{p^2 d^2}{2\hbar^2}} \cdot e^{-\frac{i}{\hbar}(px-\frac{p^2}{2m}t)}$$

Calculate the Gaussian integral of dp

$$\alpha = \frac{d^2}{2\hbar^2} + \frac{it}{2m\hbar} \text{ and } \beta = \frac{ix}{\hbar}$$

The solution is:

$$|\psi|^2 = exp\left[-\frac{x^2}{d^2} \cdot \frac{1}{1+t^2/\tau^2}\right] \text{ where } \tau = \frac{md^2}{\hbar}$$

We notice here one of the things you can do with our equation for proton-seconds is integrate from 0.5 sec to 1 sec and you get one which multiplied by the constant which is six yields six. Now look up 0.5 seconds from the data output from the program and it is magnesium, then go to one second and it is carbon, thus the integral from magnesium in time to carbon in time is carbon in protons. Now consider life as we know it is based on carbon because it has four valence electrons, but it is not based on silicon, which has four valence electrons as well, because in the presence of oxygen it readily forms SiO2 (sand or glass) leaving it unavailable to nitrogen, phosphorus, and hydrogen to make make amino acids the building blocks of life. But silicon can be doped with phosphorus, boron, gallium, and arsenic to make semiconductors -- transistor technology from which we can build artificial life (artificial intelligence, AI). We can integrate over many time ranges to explore millions of more facets to the equation:

$$\frac{1}{\alpha^2 m_p}\sqrt{\frac{h4\pi r_p^2}{Gc}}\int_{t_{Mg}}^{t_C}\frac{1}{t^2}dt = 6 = carbon(C)$$

$$t_{Mg} = 0.5 seconds$$

$$t_C = 1 second$$

We should say

$$d = \frac{Si - C}{C} = \frac{4}{3} \text{ or } d^2 = \frac{16}{9}$$

The way I am using equation 1 is $\tau = d^2$. We have:

Equation 3.14

$$|\psi|^2 = exp\left[-\frac{C^2 x^2}{(Si-C)^2} \cdot \frac{1}{1+\left[\frac{\hbar C^2}{m(Si-C)^2}\right]^2 t^2}\right]$$

Thus for hydrogen:

$$|\psi|^2 = exp\left[-\frac{9}{16}x^2 \cdot \frac{1}{1 + \frac{\hbar^2 81}{m^2 256}t^2}\right]$$

$$|\psi|^2 = (1)exp\left[-\frac{9}{16}(1proton)^2 \cdot \frac{1}{1 + \frac{(0.075)81}{(1)256}(6seconds)^2}\right] = 74\%.$$

For Helium:

$$|\psi|^2 = (2)exp\left[-\frac{9}{16}(2proton)^2 \cdot \frac{1}{1 + \frac{(0.075)81}{(4)256}(3seconds)^2}\right] = 26\%$$

This is interesting because the Universe is about 74% Hydrogen and 24% Helium, the rest of the elements making up the other 2%. Thus we can say $\hbar^2 = 0.075$ or $\hbar = 0.27386$. We have multiplied the first by 1 for Hydrogen element 1 (1 proton), and the second by 2 for helium element 2 (2 protons). In a sense then, the probabilities represent the probability of finding hydrogen and helium in the Universe. Hydrogen much of the helium were made theoretically in the Big Bang of the big bang theory. The other elements were synthesized from these by the stars. The difference between hydrogen and helium is that some of the helium was made in stars, that may be why it is 26% not 24% because it could include the remaining 2% of elements in the universe. It must be kept in mind the data we have on the universe as a whole for relative abundances can only currently be ballpark figures.

4.0 Formulation of Proton-Seconds

We can actually formulate this differently than we have. We had

$$\frac{1}{t_1}\frac{1}{\alpha^2 m_p}\sqrt{\frac{h4\pi r_p^2}{Gc}} = 6 protons$$

$$\frac{1}{t_6}\frac{1}{\alpha^2 m_p}\sqrt{\frac{h4\pi r_p^2}{Gc}} = 1 proton$$

But if t1 is not necessarily 1 second, and t6 is not necessarily six seconds, but rather t1 and t2 are lower and upper limits in an integral, then we have:

Equation 4.1 $\quad \dfrac{1}{\alpha^2 m_p}\sqrt{\dfrac{h4\pi r_p^2}{Gc}}\displaystyle\int_{t_1}^{t_2}\dfrac{1}{t^2}dt = \mathbb{N}$

This Equation is the generalized equation we can use for solving problems.

Essentially we can rigorously formulate the notion of proton-seconds by considering

Equation 4.2 $\quad \displaystyle\int_t qdt = t^2 \iint_S \rho(x,y,z)dxdy$

Is protons-seconds squared where current density is $\overrightarrow{J} = \rho \overrightarrow{v}$ and $\rho = Q/m^3$ (ρ can also be Q/m^2). We say

Equation 4.3. $\quad Q = \displaystyle\int_V \rho dV$

Keeping in mind q is not charge (coulombs) but a number of charges times seconds, here a number of protons. It is

Equation 4.4 $\quad \mathbb{N} = \dfrac{1}{\alpha^2 m_p}\sqrt{\dfrac{h4\pi r_p^2}{Gc}}$

Dividing Equation 4.2 through by by t:

Equation 4.5

$$\frac{1}{\alpha^2 m_p}\sqrt{\frac{h4\pi r_p^2}{Gc}}\int_t \frac{dt}{t} = t\iint_S \rho(x,y,z)dxdy$$

Which is proton-seconds. Dividing through by t again:

Equation 4.6
$$\frac{1}{\alpha^2 m_p} \sqrt{\frac{h 4\pi r_p^2}{Gc}} \int_t \frac{dt}{t^2} = protons$$

We see that if $\vec{J} = \rho \vec{v}$ where $\rho = Q/m^3$ and $v = m/s$ then J is l/m2 (current per square meter) is analogous to amperes per per square meter which are coulombs per second through a surface. Thus we are looking at a number of protons per second through a surface. Thus we write:

$$\frac{1}{\alpha^2 m_p} \sqrt{\frac{h 4\pi r_p^2}{Gc}} \int_{t_{Mg}}^{t_C} \frac{dt}{t^2} = 6 \int_{0.5}^{1.0} t^{-2} dt = -6(1-2) = 6$$

Is carbon where 0.5 seconds is magnesium (Mg) from the values of time corresponding to protons in the output from our program and 1.0 seconds is carbon (C). We see we have the following theorem:

Equation 4.7
$$\frac{1}{\alpha^2 m_p} \sqrt{\frac{h 4\pi r_p^2}{Gc}} \int_t \frac{dt}{t^3} = \iint_S \vec{J} \cdot d\vec{S}$$

So as an example,…

$$\frac{1}{\alpha^2 m_p} \sqrt{\frac{h 4\pi r_p^2}{Gc}} \int_{0.5}^{1.0} \frac{dt}{t^3} = \iint_S \vec{J} \cdot d\vec{S} = -3 \left(1 - \frac{1}{0.25}\right) = 9 \frac{protons}{second}$$

Is fluorine (F). Divide by xy with x=y=1 and we have current density. And multiply by 1 second which is carbon and we have protons per square meter. Charge conservation gives:

$$\frac{\partial \rho}{\partial t} + \nabla \cdot \vec{J} = 0$$

Is 9-9=0. In general…

$$\vec{J}(x, y, z) = (0,0,J) = -J\vec{k}$$

$$d\vec{S} = dx dy \vec{k}$$

$$\vec{J} \cdot d\vec{S} = (0,0,J) \cdot (0,0,dx dy) = -J dx dy$$

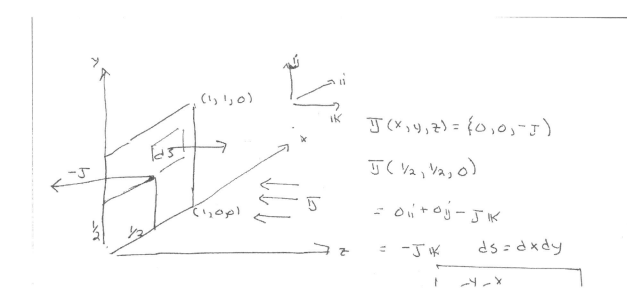

We consider the surface through which J passes a circle of radius R where J decreases linearly as r with J_0 at the center we have:

Equation 4.8

$$\iint_S \vec{J} \cdot d\vec{S} = J_0 \int_0^{2\pi} \int_0^R \left(1 - \frac{r}{R}\right) r\,dr\,d\theta = \frac{\pi J_0 R^2}{3}$$

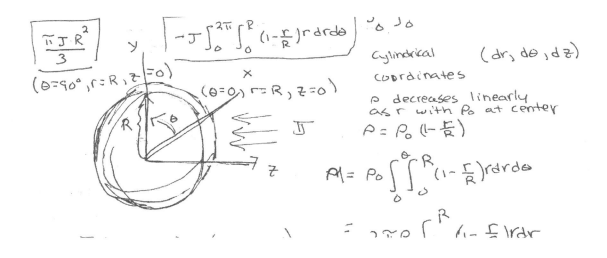

5.0 The Combined Formulations: We have

Equation 5.1 $\quad \dfrac{1}{\alpha^2 m_p} \sqrt{\dfrac{h 4 \pi r_p^2}{Gc}} \displaystyle\int_t \dfrac{dt}{t^3} = \iint_S \vec{J} \cdot d\vec{S}$

$$\dfrac{1}{\alpha^2 m_p} \sqrt{\dfrac{h 4 \pi r_p^2}{Gc}} \int_{t_c}^{t_H} \dfrac{dt}{t^3} = \iint_S \vec{J} \cdot d\vec{S} = -3 \left(\dfrac{1}{36} - \dfrac{1}{1} \right) = \dfrac{35}{12} \cdot \dfrac{protons}{second}$$

$$x^2 = \dfrac{35}{12} \cdot \dfrac{protons}{second}$$

Let's say J is carbon and six protons and that that is over one second and one meter, then...

$$6 \cdot \dfrac{protons}{second \cdot m^2} \cdot x^2 = \dfrac{35}{12}$$

$$x^2 = \dfrac{35}{72} m^2 \approx 0.5 m^2 = \dfrac{1}{2} m^2$$

$$x = 0.6972 m \approx 0.7 m = \dfrac{7}{10} m$$

Now consider Giordano's Formulation:

Equation 5.2 $\quad h(1 + \alpha) N_A \mathbb{H} = 6Gx$

It implies

$$x = 1.00 kg^2 \dfrac{s}{m}$$

We should change x to some other letter, like k, since we want to use x in J, the flow of protons per second per square meter. Then,...

$$k = 1.00 kg^2 \dfrac{s}{m}$$

Combining this with our equation for proton seconds

Equation 5.3 $\quad 1.00 kg^2 \dfrac{s}{m} \cdot \dfrac{1}{6} \cdot \dfrac{7}{10} m \cdot \alpha^2 m_p \sqrt{\dfrac{Gc}{h 4 \pi r_p^2}} = \dfrac{21}{5} kg^2 = \dfrac{7}{60} kg^2$

$m = 0.34 kg$ (as the mass for six protons per square meter per second)

While we have masses characteristic of the microcosmos like protons, and masses characteristic of the macrocosmos, like the minimum mass for a star to become a neutron star as opposed to a white dwarf after she novas (The Chandrasekhar limit) which is 1.44 solar masses, we do not have a characteristic mass of the intermediary world where we exist, a truck weighs several tons and tennis ball maybe around a hundred grams. To find that mass let us take the geometric mean between the mass of a proton and the mass of 1.44 solar masses. We could take the average, or the harmonic mean, but the geometric mean is the squaring of the proportions, it is the side of a square with the area equal to the area of the rectangle with these proportions as its sides. We have:

$$M_\odot = 1.98847E30 kg$$

We multiply this by 1.44 to get 2.8634E30kg. The mass of a proton is $m_p = 1.67262E - 27kg$. We have the intermediary mass is:

$$m_i = \sqrt{(2.8634E30)(1.67262E - 27)} = 69.205 kg$$

$$\frac{69.205}{0.34} = 203.544$$

But let us approach these precisely in terms of concept:

Equation 5.4

$$Jx^2 \cdot \frac{h(1 + \alpha)N_A \cdot \mathbb{E}}{6Gc} = \frac{35}{12} \cdot \frac{protons}{second} \left(1.00 kg^2 \frac{s}{m}\right)$$

Equation 5.5

$$Jx^2 \frac{1}{t_1} \cdot \frac{h(1 + \alpha)N_A \cdot \mathbb{E}}{6Gc} = \frac{35}{12} \cdot \frac{protons}{second} \left(1.00 \frac{kg^2}{m}\right)$$

Where t1=1second. And putting in the intermediary mass,…

Equation 5.6

$$\frac{Jx^2}{m_i^2 t_1} \cdot \frac{h(1 + \alpha)N_A \cdot \mathbb{E}}{6Gc} = \frac{35}{12} \cdot \frac{protons}{second} \left(1.00 m^{-1}\right) = 6E - 4 \frac{protons}{meter \cdot second}$$

That is:

On average the solar system contains 5 atoms in cubic cm. Interstellar space about one.

Since:

$$Jx^2 = \frac{1}{t^2} \frac{1}{\alpha^2 m_p} \sqrt{\frac{h 4\pi r_p^2}{Gc}}$$

Equation 6 becomes:

Equation 5.7

$$\frac{1}{\alpha^2 m_p}\sqrt{\frac{h4\pi r_p^2}{Gc}}\frac{1}{m_i^2 t_1^3}\cdot\frac{h(1+\alpha)N_A\cdot\mathbb{E}}{6Gc}=6E-4\frac{protons}{meter\cdot second}$$

We see then, combining my proton-seconds with Giordano's relationship produces protons flowing in a line. Where with my proton-seconds we had a surface integral, combining it with Giordano's relationship we have a line integral. With proton-seconds we had protons flowing through a surface but here we have them traveling in a line. which is the divergence theorem. Since

$$\frac{1}{\alpha^2 m_p}\sqrt{\frac{h4\pi r_p^2}{Gc}}$$

Is six, and…

$$\frac{h(1+\alpha)N_A\cdot\mathbb{E}}{6Gc}$$

Has a six in the denominator, the sixes cancel, except in units. But we know the overall resulting units so we have for the value,…

$$\frac{1}{m_i^2}=2E-4$$

But we have to multiply that by 35/12 and we have our value 6E-4. Everything checks out. We just need to verify the units…

$$(proton-seconds)\frac{1}{kg^2\cdot seconds^3}\cdot kg^2\cdot\frac{sec}{meter}=\frac{protons}{meter\cdot second}$$

If the vector field is an inverse square law, then it is a conservative vector field because its curl is zero:

$$\nabla\times\overrightarrow{F}=\begin{pmatrix}e_r & e_\theta r & e_\phi rsin\theta\\ \frac{\partial}{\partial r} & \frac{\partial}{\partial\theta} & \frac{\partial}{\partial\phi}\\ F_r & rF_\theta & rF_\phi sin\theta\end{pmatrix}=0$$

We would then proceed to apply the gradient theorem,…

$$\int_C \nabla\phi \cdot d\vec{r} = \int_b^a d\phi = \phi(\vec{r_2}) - \phi(\vec{r_1})$$

We want to find the scalar field ϕ such that $\vec{J} = \nabla\phi$.

...But first the divergence theorem. The divergence theorem is:

$$\int_V (\nabla \cdot \vec{u})dV = \oint_S \vec{u} \cdot d\vec{S}$$

For a sphere,...

$$r^2 = x^2 + y^2 + z^2$$

$$r = (x^2 + y^2 + z^2)^{1/2}$$

$$\nabla = \frac{\partial}{\partial x}\vec{i} + \frac{\partial}{\partial y}\vec{i} + \frac{\partial}{\partial z}\vec{z}$$

$$\vec{u} = \frac{x\vec{i} + y\vec{j} + z\vec{k}}{(x^2 + y^2 + z^2)^{3/2}} = \frac{\vec{r}}{(x^2 + y^2 + z^2)^{3/2}} = \frac{\hat{r}}{r^2} = \frac{\vec{r}}{r^3}$$

$$J_x = \frac{\partial\vec{u}}{\partial x} = \frac{\partial}{\partial x}\frac{x}{(x^2 + y^2 + z^2)^{3/2}} = \frac{-2x^2 + y^2 + z^2}{(x^2 + y^2 + z^2)^{5/2}}$$

$$J_y = \frac{\partial\vec{u}}{\partial y} = \frac{\partial}{\partial y}\frac{y}{(x^2 + y^2 + z^2)^{3/2}} = \frac{x^2 - 2y^2 + z^2}{(x^2 + y^2 + z^2)^{5/2}}$$

$$J_z = \frac{\partial\vec{u}}{\partial z} = \frac{\partial}{\partial z}\frac{z}{(x^2 + y^2 + z^2)^{3/2}} = \frac{x^2 + y^2 - 2z^2}{(x^2 + y^2 + z^2)^{5/2}}$$

$$\nabla \cdot \phi =$$

$$\left(\frac{\partial}{\partial x}\vec{i} + \frac{\partial}{\partial y}\vec{i} + \frac{\partial}{\partial z}\vec{z}\right) \cdot \left(\frac{x}{(x^2 + y^2 + z^2)^{1/2}}\vec{i} + \frac{y}{(x^2 + y^2 + z^2)^{1/2}}\vec{j} + \frac{z}{(x^2 + y^2 + z^2)^{1/2}}\vec{k}\right)$$

$$= \frac{-2x^2 + y^2 + z^2}{(x^2 + y^2 + z^2)^{5/2}} + \frac{x^2 - 2y^2 + z^2}{(x^2 + y^2 + z^2)^{5/2}} + \frac{x^2 + y^2 - 2z^2}{(x^2 + y^2 + z^2)^{5/2}} = 0$$

Because $-2x^2 + x^2 + x^2 = 0$, as well for the sum of y terms and for z terms. This is one of Maxwell's Equations, the one for the magnetic or $\vec{B}$ field which is:

$$\nabla \cdot B = 0$$

We could proceed like this in rectangular coordinates, but we would be off our rocker; we choose to exploit the spherical symmetry and the problem merely becomes in spherical coordinates (I just provided it like this because it illuminates the nature of the geometry)...

Which states that there can be no magnetic monopoles. We could actually do this quite simply in spherical coordinates.

$$\nabla \cdot \vec{A} = \frac{1}{r^2}\frac{\partial(r^2 A_r)}{\partial r} + \frac{1}{rsin\theta}\frac{\partial}{\partial\theta}(A_\phi sin\theta) + \frac{1}{rsin\theta}\frac{\partial A_\phi}{\partial\phi}$$

If $A_r = \frac{1}{r^2}$, then since A_θ and A_ϕ are zero in an inverse square field, we have in the first term the derivative of a constant which is zero. Thus if we have that

$$\nabla \cdot \vec{J} = 0$$

Then a field that satisfies this is an inverse square field. The gradient of an inverse square field is:

$$\nabla \vec{J} = \left(\frac{-2x^2 + y^2 + z^2}{(x^2 + y^2 + z^2)^{5/2}}, \frac{x^2 - 2y^2 + z^2}{(x^2 + y^2 + z^2)^{5/2}}, \frac{x^2 + y^2 - 2z^2}{(x^2 + y^2 + z^2)^{5/2}} \right)$$

We can proceed in spherical coordinates:

$$\nabla f(r, \theta, \phi) = \frac{\partial f}{\partial r}\vec{e_r} + \frac{1}{r}\frac{\partial f}{\partial\theta}\vec{e_\theta} + \frac{1}{rsin\theta}\frac{\partial f}{\partial\phi}e_\phi$$

The $\frac{\partial f}{\partial\theta}$ is zero as well the $\frac{\partial f}{\partial\phi}$. We simply have

$$\frac{\partial \vec{J}}{\partial r}\vec{e_r} = -\frac{\hat{r}}{r^3}$$

$$\frac{\partial}{\partial r}\frac{1}{r} = \frac{1}{r^2}$$

Meaning

$$\vec{J} = \nabla\phi = \frac{J_0}{r^2}$$

Is an inverse square field. But let us return to the divergence theorem,

$$\int_V (\nabla \cdot \vec{u})dV = \oint_S \vec{u} \cdot d\vec{S}$$

We see the volume integral (which we just did) is equal to the surface integral by way of the vector field $\overrightarrow{u}$, because in the former we take the divergence of it which is the three dimensional analog of a derivative, thus we have a volume integral (a triple integral) of a derivative equal to a surface integral (a double integral) which is the analog of a integral of a derivative equal to a function which means this is the analog of the fundamental theorem of calculus but in three dimensional space. So where the fundamental theorem of calculus allows us to solve differential equations, the divergence theorem allows us to solve partial differential equations. But as well it relates the divergence of a vector field in a volume to the flux of that vector field passing through a surface. Thus if we want to solve one of these two we can either do it directly by brute force, for instance compute the volume integral directly, or solve it by computing the surface integral, and vice versa. You have to choose the one that is easier to solve. We just did the volume integral. But, let us do it by solving the surface integral, because it can be illuminating.

$$\overrightarrow{J} = \nabla\phi = \frac{J_0}{r^2}$$

The surface $\overrightarrow{S}$ is a sphere. Thus $d\overrightarrow{S}$ is

$$d\overrightarrow{S} = r\hat{r}R^2\sin\theta d\theta d\phi$$

Of course the surface element is r directed away from the surface of the sphere, but we have to multiply by $R^2\sin\theta d\theta d\phi$ when going from rectangular coordinates, the transformation which is given by the Jacobean matrix, but that is a subject for linear algebra, but is:

$$dxdydz = \left|\left|\begin{pmatrix} \frac{\partial x}{\partial r} & \frac{\partial x}{\partial \theta} & \frac{\partial x}{\partial \phi} \\ \frac{\partial y}{\partial r} & \frac{\partial y}{\partial \theta} & \frac{\partial y}{\partial \phi} \\ \frac{\partial z}{\partial r} & \frac{\partial z}{\partial \theta} & \frac{\partial z}{\partial \phi} \end{pmatrix}\right|\right| drd\theta d\phi$$

Which means a volume element dV is

$$dV = dxdydz = r^2\sin\theta drd\theta d\phi$$

And a surface element dS is

$$dS = r^2\sin\theta d\theta d\phi$$

$$\overrightarrow{u} \cdot d\overrightarrow{S} = \frac{1}{r^2}R^2\sin\theta drd\theta d\phi$$

$$-\frac{1}{R^2}R^2\cos(0°)(2\pi) = 2\pi R$$

If

$$\int \nabla \cdot \vec{u} \, dV = 2\pi R$$

Which is a constant C, then

$$\nabla \cdot \vec{u} = 0$$

Because

$$\int 0 \, dV = constant$$

Back to the problem at hand. In equation 1:

$$\frac{1}{\alpha^2 m_p} \sqrt{\frac{h 4\pi r_p^2}{Gc}} \int_t \frac{dt}{t^3} = \iint_S \vec{J} \cdot d\vec{S}$$

J is protons per square meter per second, but since combined with Giordano's relationship is equation 7:

$$\frac{1}{\alpha^2 m_p} \sqrt{\frac{h 4\pi r_p^2}{Gc}} \frac{1}{m_i^2 t_1^3} \cdot \frac{h(1+\alpha)N_A \cdot \mathbb{E}}{6Gc} = 6E - 4 \frac{protons}{meter \cdot second}$$

J is protons per meter per second. And we have come to:

$$\vec{J} = \nabla \phi = \frac{J_0}{r^2}$$

Where

$$J_0 = 6E - 4 \frac{protons}{meter \cdot second}$$

This gives

$$J = \frac{proton \cdot meters}{second}$$

If r is the radius of the Earth, R_E. Then,..

$$\frac{6.09E - 4}{6,371,000^2} = 1.500E - 17 \frac{proton \cdot meters}{second}$$

Is a form of momentum.

Since we have Maxwell's equation for the magnetic field is

$$\nabla \cdot B = 0$$

What about Maxwell's equation for the electric field? Its divergence is not zero but is a constant (Gauss' Equation):

$$\nabla \cdot \overrightarrow{E} = \frac{\rho}{\epsilon_0}$$

Where ρ is the total electric charge density, and $\epsilon_0 = 8.8541878128E - 12 F/m$ (farads per meter). The force between two charges is:

$$F = \frac{1}{4\pi\epsilon_0} \frac{q_1 q_2}{r^2}$$

Is Coulomb's law. We have then that

$$\nabla \cdot \overrightarrow{J} = C$$

Some constant C. In earlier work I had addressed this area from what is in this paper in section **2.0 The Periodic Table of the Elements**. Lets address that in a new section:

6.0 The Halfwave

By making the approximation

$$\frac{2SiGe}{Si + Ge} \approx Ge - Si$$

In

$$\frac{Si(As - Ga)}{B} + \frac{Ge(P - Al)}{B} = \frac{2SiGe}{Si + Ge}$$

We have

$$Si\frac{\Delta Ge}{\Delta S} + Ge\frac{\Delta Si}{\Delta S} = B$$

$\Delta Si = P - Al$ is the differential across Si, $\Delta Ge = As - Ga$ is the differential across Ge and $\Delta S = Ge - Si$ is the vertical differential.

Which is Ampere's Circuit Law

$$\nabla \times \overrightarrow{B} = \mu_0 \overrightarrow{J} + \mu_0\epsilon\frac{\partial \overrightarrow{E}}{\partial t}$$

We see if written

$$Si\frac{\Delta Ge}{\Delta S} = B - Ge\frac{\Delta Si}{\Delta S}$$

Which is interesting because it is semiconductor elements by molar mass which are used to make circuits.

We say Φ (Phi) is given by

$$a = b + c \text{ and } \frac{a}{b} = \frac{b}{c}$$

And

$$\Phi = a/b = 1.618$$

$$\phi = b/a = 0.618$$

ϕ (phi) the golden ratio conjugate. We also find

$$(\phi)\Delta Ge + (\Phi)\Delta Si = B$$

Thus since

$$\nabla \times \overrightarrow{B} = \mu \overrightarrow{J} + \mu\epsilon_0\frac{\partial\overrightarrow{E}}{\partial t}$$

$$Si\frac{\Delta Ge}{\Delta S} = B - Ge\frac{\Delta Si}{\Delta S}$$

And we have

$$\Delta Ge = \frac{\Delta S}{Si}B - \frac{Ge}{Si}\Delta Si$$

$$\left(\nabla^2 - \frac{1}{c^2}\frac{\partial^2}{\partial t}\right)\overrightarrow{E} = 0$$

$$\left(\nabla^2 - \frac{1}{c^2}\frac{\partial^2}{\partial t}\right)\overrightarrow{B} = 0$$

$$c = \frac{1}{\sqrt{\epsilon_0\mu}} \approx \phi$$

We see μ and ϵ_0 are both Φ and c is ϕ in the Si (silicon) field wave, but for E and B fields c is the speed of light.

$$\epsilon_0 = 8.854E - 12F \cdot m^{-1}$$

$$\mu = 1.256E - 6H/m$$

$$\frac{Ge}{Si} = \mu\epsilon_0$$

$$\frac{\Delta S}{Si} = \mu$$

$$\left(\nabla^2 - \frac{1}{\phi^2}\frac{\partial^2}{\partial x}\right)\overrightarrow{Si} = 0$$

$$\left(\nabla^2 - \frac{1}{\phi^2}\frac{\partial^2}{\partial x}\right)\overrightarrow{Ge} = 0$$

To find the Si wave our differentials are

$$\Delta C = N - B = 14.01 - 10.81 = 3.2$$

$$\Delta Si = P - Al = 30.97 - 26.98 = 3.99$$

$$\Delta Ge = As - Ga = 74.92 - 69.72 = 5.2$$

$$\Delta Sn = Bi - In = 121.75 - 114.82 = 6.93$$

$$\Delta Pb = Bi - Tl = 208.98 - 204.38 = 4.6$$

It is amazing how accurately we can fit these differentials with an exponential equation for the upward increase. The equation is

$$y(x) = e^{0.4x} + 1.7$$

$$y(x) = e^{\frac{2}{5}x} + \frac{17}{10}$$

This is the halfwave:

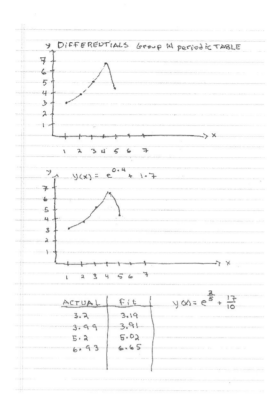

$$y(x) = e^{0.4x} + 1.7$$

$$y(x) = e^{\frac{2}{5}x} + \frac{17}{10}$$

$$y(x) = e^{\frac{B}{Al}x} + \frac{Ag}{Cu}$$

$$\frac{B}{Al} = \frac{10.81}{26.98} = 0.400667$$

$$\frac{Ag}{Cu} = \frac{107.87}{63.55} = 1.6974 \approx 1.7$$

This is pertinent to section 5.0 because we are looking at:

$$\nabla \cdot \vec{E} = \frac{\rho}{\epsilon_0}$$

And we have found here:

$$c = \frac{1}{\sqrt{\epsilon_0 \mu}} \approx \phi$$

7.0 Philosophy: When we say there are 6E-14 protons per meter per second we are saying

$$\frac{\frac{protons}{meter}}{second}$$

But this is

$$\frac{protons}{meter} \cdot \frac{1}{second} = \frac{protons}{meter \cdot second}$$

However our statement can result in the inverse deepening on how we **interpret** our statement, in that

$$protons \cdot \frac{1}{meters/second} = \frac{protons \cdot second}{meter}$$

It is interesting because 6 seconds **corresponds** to 1 proton and we can write this as 1 proton over 6 seconds:

$$proton \cdot seconds$$

Or, 1 proton for every 6 seconds:

$$\frac{protons}{6 seconds}$$

As if the inverse were the same thing. Either one **works.** while both can lead to the same **discovery**, the approaches in getting there are different. The discovery is **useful** but any meaning is lost. This is the **elusiveness** of truth. Whether we write

$$proton \cdot seconds$$

Or,...

$$\frac{protons}{seconds}$$

Is not important; what **transcends** this is **geometry.** We can say six equal radius spheres, which are protons closest packed as the regular hexagon as drawn here:

The regular hexagon is an equal sided, equal angled polygon with six sides and 60 degree coordination as in the following drawings:

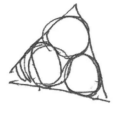 

Which allows us to approximate pi to its closest integer as in the following drawings, because

$$\frac{S_1 + S_2 + S_3 + S_4 + S_5 + S_6}{2r} = 3 \approx \pi$$

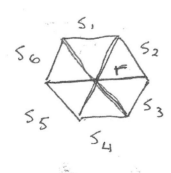

7.0 Building A Matrix

We pull these Al elements out of the periodic table of the elements to make an Al periodic table:

	13	14	15
2	B		
3	Al	Si	P
4	Ga	Ge	As

We now notice we can make a 3 by 3 matrix of it, which lends itself to to the curl of a vector field by including biological elements carbon C (above Si):

$$\begin{pmatrix} \vec{i} & \vec{j} & \vec{k} \\ \dfrac{\partial}{\partial x} & \dfrac{\partial}{\partial y} & \dfrac{\partial}{\partial z} \\ (-C \cdot P)y & (Si \cdot Ga)z & (Ge \cdot As)y \end{pmatrix} =$$

$$(Ge \cdot As - Si \cdot Ga)\vec{i} + (C \cdot P)\vec{k} =$$

$$\left[(72.64)(74.92) - (28.09)(69.72) \right] \vec{i} + \left[(12.01)(30.97) \right] \vec{k}$$

Which resulted in Stokes theorem (Beardsley, Essays In Cosmic Archaeology Volume 3):

$$\sqrt[5]{\int_{Si}^{Ge} \int_{Si}^{Ge} \nabla \times \vec{u} \cdot d\vec{a}} = exp\left(\frac{1}{Ge - Si} \int_{Si}^{Ge} ln(x)dx \right)$$

Where

$$\nabla \times \vec{u} = (Ge \cdot As - Si \cdot Ga)\vec{i} + (C \cdot P)\vec{k}$$

$$d\vec{a} = \left(zdydz\vec{i} + ydydz\vec{k} \right)$$

$$\vec{u} = (-C \cdot P)y\vec{i} + (Si \cdot Ge)z\vec{j} + (Ga \cdot As)y\vec{k}$$

We were then able to write this with product notation

$$\sqrt[5]{\int_{Si}^{Ge}\int_{Si}^{Ge}\nabla\times\overrightarrow{u}\cdot d\overrightarrow{a}}=\sqrt[n]{\prod_{i=1}^{n}x_i}$$

While we have the Al BioMatrix

B. C. N.

Al. Si. P.

Ga. Ge. As.

Which we used to formulate a similar equation (Beardsley, Essays In Cosmic Archaeology Volume 2). We can form another 3X3 matrix we will call the electronics matrix (Beardsley, Cosmic Archaeology, Volume Three):

Ni. Cu. Zn.

Pd. Ag. Cd.

Pt. Au. Hg.

We can remove the 5th root sign in the above equation by noticing

$$\prod_{i=1}^{5}x_i = Si\cdot Ge\cdot Cu\cdot Ag\cdot Au$$

$$=(28.085)(72.64)(12.085)(107.8682)(196.9657)=$$

$$523,818,646.5\frac{g^5}{mol^5}$$

Where we have substituted carbon (C=12.01) the core biological element for copper (Cu).

But since we have:

$$\int_{Si}^{Ge}\int_{Si}^{Ge}(\nabla\times\overrightarrow{u})\cdot d\overrightarrow{a}=170,535,359.662(g/mol)^5$$

We take the ratio and have

$$\frac{523,818,646.5}{170,535,359.662} = 3.0716$$

Almost exactly 3 which is the ratio of the perimeter of regular hexagon to its diameter used to estimate pi in ancient times by inscribing it in a circle:

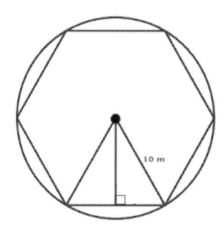

Perimeter=6

Diameter=2

6/2=3

$$\pi = 3.141...$$

Thus we have the following equation…

Equation 7.1.
$$\pi \int_{Si}^{Ge} \int_{Si}^{Ge} (\nabla \times \overrightarrow{u}) \cdot d\overrightarrow{a} = \prod_{i=1}^{5} x_i$$

Showing the calculation using the most accurate data possible…

Ge=72.64
As=74.9216
Si=28.085
Ga=69.723
C=12.011
P=30.97376200

$$(Ge \cdot As - Si \cdot Ga)\overrightarrow{i} + (C \cdot P)\overrightarrow{k} =$$

$$\left[(72.64)(74.9216) - (28.085)(69.723)\right]\overrightarrow{i} + \left[(12.011)(30.97376200)\right]\overrightarrow{k} =$$

$$3{,}484.134569\left(\frac{g}{mol}\right)^2\vec{i}+372.025855\left(\frac{g}{mol}\right)^2\vec{k}$$

$$\int_{Si}^{Ge}\int_{Si}^{Ge}\left(3{,}484.134569\left(\frac{g}{mol}\right)^2\vec{i}+372.025855\left(\frac{g}{mol}\right)^2\vec{k}\right)\cdot\left(zdydz\vec{i}+ydydz\vec{k}\right)$$

$$\int_{Si}^{Ge}\int_{Si}^{Ge}\left(3{,}484.134569\left(\frac{g}{mol}\right)^2\cdot zdzdy+372.025855\left(\frac{g}{mol}\right)^2\cdot ydzdy\right)$$

$$\int_{Si}^{Ge}3{,}484.134569\left(\frac{(72.64-28.085)^2}{2}\right)dy+\int_{Si}^{Ge}372.025855y\cdot(72.64-28.085)dy$$

$$3458261.42924\left(\frac{g}{mol}\right)^4(72.64-28.085)+16575.6119695\left(\frac{g}{mol}\right)^3\left(\frac{(72.64-28.085)^2}{2}\right)$$

=154,082,837.980+16,452,521.6822=

$$170{,}535{,}359.662\left(\frac{g}{mol}\right)^5$$

$$\prod_{i=1}^{5}x_i=Si\cdot Ge\cdot C\cdot Ag\cdot Au=$$

(28.085)(72.64)(12.085)(107.8682)(196.9657)=

$$523{,}818{,}646.5\frac{g^5}{mol^5}$$

Where we have substituted carbon C=12.01 for copper Cu. We use Cu, Ag, Au because they are the middle column of our electronics matrix, they are the finest conductors used for electrical wire. We use C, Si, Ge because they are the middle column of our AI Biomatrix. Si and Ge are the primary semiconductor elements used in transistor technology (Artificial Intelligence) and C is the core element of biological life. We have

$$\frac{523{,}818{,}646.5}{170{,}535{,}359.662}=3.0716$$

$$\pi=3.141...$$

Perimeter/Diameter of regular hexagon = 3.00

$$\frac{3.141 + 3.00}{2} = 3.0705$$

The same value as our 3.0716 if taken at two places after the decimal.

Thus we are interested in:

Equation 7.1 $\qquad \pi \int_{Si}^{Ge} \int_{Si}^{Ge} (\nabla \times \vec{u}) \cdot d\vec{a} = \prod_{i=1}^{5} x_i$

Equation 7.2 $\qquad \prod_{i=1}^{5} x_i = Si \cdot Ge \cdot C \cdot Ag \cdot Au$

Conclusion

Section 2.0

The AI elements are formulations of geometry, namely Stoke's Theorem in the harmonic mean:

$$\int_0^1 \int_0^1 \frac{Si}{B}(As - Ga)dydz = \frac{1}{3}\int_{Si}^{Ge} dz$$

$$\int_0^1 \int_0^1 \frac{Ge}{B}(P - Al)dydz = \frac{2}{3}\int_{Si}^{Ge} dz$$

Harmonic mean form:

$$\frac{Si}{B}(As - Ga) + \frac{Ge}{B}(P - Al) = \frac{Ge - Si}{\int_{Si}^{Ge} \frac{dx}{x}}$$

Section 3.0

Not only is carbon the element by which we historically developed our definitions of molar mass, we see here it defines inertia with the unit of the second which we have shown has its foundation in the vector equilibrium:

$$\frac{1}{t_1 \alpha^2 m_p}\sqrt{\frac{h4\pi r_p^2}{Gc}} = 6 protons$$

$$\frac{1}{6\alpha^2 m_p}\sqrt{\frac{h4\pi r_p^2}{Gc}} = 1.004996352 seconds$$

This not only defines the the periodic table of the elements as periodic over 18 groups:

$$\frac{3}{\alpha^2 m_p}\sqrt{\frac{h4\pi r_p^2}{Gc}}\int_{t_1}^{t_2} \frac{1}{t^2}dt = 18$$

But predicts the radius of a proton as founded on carbon because we have found carbon's time representative to be 1 second:

$$r_p = k\alpha^2 m_p \sqrt{\frac{Gc}{4\pi h}}$$

$$r_p = 8.288587E - 16m = 0.829 fm$$

Where the experimental value is given by:

$$r_p = 0.833 +/- 0.014$$

A very interesting thing here is looking at the values generated by the program, the smallest integer value, 1 second produces 6 protons (carbon) and the largest integer value 6 seconds produces one proton (hydrogen). Beyond six seconds you have fractional protons, and the rest of the elements heavier than carbon are formed by fractional seconds. These are the hydrocarbons the backbones biological chemistry.

$$\frac{1}{\alpha^2 m_p}\sqrt{\frac{h4\pi r_p^2}{Gc}} = (6 protons)(1 second)$$

$$\frac{1}{\alpha^2 m_p}\sqrt{\frac{h4\pi r_p^2}{Gc}} = (1 proton)(6 seconds)$$

By way of Giordano's Relationship we see

$$h\frac{(1+\alpha)}{G} \cdot N_A \mathbb{H} = 6.0003 kg^2 \cdot \frac{s}{m}$$

We have a pattern throughout the periodic table as related to G for any element:

$$N_A H = 6.02E23\frac{atoms}{gram} \cdot \frac{1 gram}{atom} = 6.02E23$$

$$h\frac{(1+\alpha)}{G} \cdot N_A \mathbb{H} = 6.0003 kg^2 \cdot \frac{s}{m}$$

$$N_A \cdot \mathbb{E} = 6E23$$

We have a definition for the radius of the solar system through the Al elements

$$R_s = \sqrt{\frac{3M_p}{\pi(0.78Ge + 0.21Si)}}$$

And can predict the relative abundances of Hydrogen and Helium from which the other elements are made in the stars:

Hydrogen

$$|\psi|^2 = (1)exp\left[-\frac{9}{16}(1proton)^2 \cdot \frac{1}{1 + \frac{(0.075)81}{(1)256}(6seconds)^2}\right]$$

Helium

$$|\psi|^2 = (2)exp\left[-\frac{9}{16}(2proton)^2 \cdot \frac{1}{1 + \frac{(0.075)81}{(4)256}(3seconds)^2}\right]$$

Section 6.0

We have a halfwave

$$y(x) = e^{\frac{B}{Al}x} + \frac{Ag}{Cu}$$

In accordance with

$$\frac{Si(As - Ga)}{B} + \frac{Ge(P - Al)}{B} = \frac{2SiGe}{Si + Ge}$$

That determines the role of the speed of light in our geometric formulation of the elements is in the golden ratio:

$$c = \frac{1}{\sqrt{\epsilon_0 \mu}} \approx \phi$$

Section 7.0

And we have a geometric formulation for the electronics elements that is in the geometric mean as the Al elements were in the harmonic mean.

$$\pi \int_{Si}^{Ge} \int_{Si}^{Ge} (\nabla \times \vec{u}) \cdot d\vec{a} = \prod_{i=1}^{5} x_i$$

$$\prod_{i=1}^{5} x_i = Si \cdot Ge \cdot C \cdot Ag \cdot Au$$

The geometric mean is best described by product calculus hence this formulation with the symbol

$$\prod_{i=0}^{n}$$

Which is a calculus that has not become standard university curriculum though purports that calculus is actually a subset of it.

The Author

48445733R00033